Anne Raquel Sotiles
Vanderlei A. de Lima
Larissa M.dos Santos Tonial

Readjusting experimental protocols in Chemistry

Anne Raquel Sotiles
Vanderlei A. de Lima
Larissa M.dos Santos Tonial

Readjusting experimental protocols in Chemistry

Reducing waste

ScienciaScripts

Imprint
Any brand names and product names mentioned in this book are subject to trademark, brand or patent protection and are trademarks or registered trademarks of their respective holders. The use of brand names, product names, common names, trade names, product descriptions etc. even without a particular marking in this work is in no way to be construed to mean that such names may be regarded as unrestricted in respect of trademark and brand protection legislation and could thus be used by anyone.

Cover image: www.ingimage.com

This book is a translation from the original published under ISBN 978-3-330-77161-1.

Publisher:
Sciencia Scripts
is a trademark of
Dodo Books Indian Ocean Ltd. and OmniScriptum S.R.L publishing group

120 High Road, East Finchley, London, N2 9ED, United Kingdom
Str. Armeneasca 28/1, office 1, Chisinau MD-2012, Republic of Moldova, Europe
Managing Directors: Ieva Konstantinova, Victoria Ursu
info@omniscriptum.com

Printed at: see last page
ISBN: 978-620-8-63827-6

CONTENTS

INTRODUCTION

Chemistry is the science that studies the composition and properties of the substances that make up the entire universe. It is known that most of the discoveries that revolve around chemical processes and reactions were made through experiments, and so chemistry is a science that requires experience in the laboratory.

When experiments are carried out, waste is generated from leftover samples and liquid or solid products from chemical processes, as well as gases and vapors from reactions. Therefore, the generation of chemical waste should be reduced to a minimum by requesting only the necessary number of samples to be used and, whenever possible, adopting methods that use the minimum number of samples (FERRAZ; FEITOSA, 2003). It is in this sense that the concept of Green Chemistry arises, which, according to Prado (2003), is the creation, development and application of chemical products and processes to reduce or eliminate the use and generation of toxic substances. Green Chemistry is part of Environmental Chemistry, which studies phenomena occurring in nature, whether due to anthropogenic action or not.

Today's capitalist society and excessive consumerism are responsible for increasing environmental degradation and have shown the limits of nature, such as environmental imbalances. By implementing the 3Rs - Reduce, Reuse and Recycle - it is possible to become ecologically aware and take care of the environment in order to achieve sustainable development.

With the aim of reducing the amount of reagents needed for reactions to take place in a visible way, this work was developed, in which some practical classes that are currently carried out in the subject of General Chemistry in the Integrated High School of a school in a municipality in the southwest of Paranà, were analyzed, tested and reformulated, so that the generation of waste was also minimized. With the same objective in mind, new experimental protocols were also tested, adapting their methodologies, with the aim of collaborating and expanding new perspectives in Green Chemistry.

TEACHING CHEMISTRY AND EXPERIMENTATION

Chemistry is the science that studies matter and the changes it undergoes in macroscopic and microscopic form, as well as describing the phenomena that occur (ATKINS, P.; JONES, L., 2012).

The very essence of chemistry reveals the importance of introducing this type of activity to the student environment. This science relates to nature, so experiments give students a more scientific understanding of the transformations that occur in nature, as well as their modification. Teaching chemistry should develop students' ability to understand the chemical phenomena present in their daily lives (FARIAS, 2009).

According to Giordan (1999), experimentation serves not only as an instrument for developing skills, but also as a vehicle for legitimizing scientific knowledge, insofar as the data extracted from experiments is the final word on understanding the phenomenon in question.

The concept of practical activity cannot be limited only to those that are created and reproduced in the classroom or laboratory, but also materialized in social experience and which permeate the negotiation of meaning from the students' point of view. From this point of view, socio-environmental issues play a crucial role, in that they provide an individual perception that motivates a collective conscience, which can result in changes in attitudes towards the concept of the environment, making experimentation an environmental education action (SILVA, 2008).

Universities are institutions where scientific knowledge is produced, where new products and technologies usually emerge, building their role in society and directly interfering in it.

A student's training goes far beyond the execution of a particular course , the experiences through contact with teachers, laboratories, learning, basic scientific research activities, among others. There's no denying that this experience will be part of this future education professional's life, with possible repercussions in our society. Universities, as opinion formers, must provide opportunities to create conditions for self-evaluation, seeking to train future professionals with environmental awareness, ethics and co-responsibility. In this sense, it is also important to emphasize the fact that there is no sustainability in the socio-environmental structure of universities, whether in the relationship between people, or the relationship between nature, people and waste (LIMA, 2011).

The media often relates the word "chemistry" to factors that are harmful to health and there is public concern about toxic chemical products, and environmental damage is a reality that cannot be hidden (BAIRD, 2002; FARIAS, 2009). Therefore, the current challenge for chemistry is to transform its old processes into more economical and cleaner ones, making chemistry more positive for public opinion. Chemical research must be developed with the concern of minimizing not only the risks of processes and effluents, but also the guarantee of quality of chemical products for the consumer (FARIAS, 2009).

For many years, university chemistry teaching and research laboratories have had to deal with the waste they produce, since experimental activities in the field of chemistry are usually risky and potentially polluting. Chemistry laboratories generate waste from the remains of liquid or solid products from chemical processes, as well as gases and vapors from reactions (FERRAZ; FEITOZA, 2004; SILVA; MACHADO, 2008; GILONI-LIMA, LIMA, KATAOKA, 2011).

Educational institutions are responsible for the training of their students and, consequently, for their behavior as citizens of the world. Therefore, these institutions must be aware of the problem of waste and the lack of waste management, advocating that activities should follow good laboratory practices. In addition, they should focus on recycling programs and not generating waste (TEIXEIRA; RIBEIRO, 2015).

According to Lenardâo et al. (2003) apud SILVA (2008), although the generation of waste and all the implications linked to it permeate the educational environment, there has still been no concrete incorporation into the daily actions of educational institutions. This can be seen in the majority of experimental practices, whether in teaching, research or extension, which are not guided by the guiding principles of Green Chemistry.

Society's current awareness of environmental issues, the process of globalization and increasing regulations confirm the need to create a waste management program in educational institutions. The management and safe disposal of laboratory waste sets a good example to students and trains them to be responsible citizens and scientists. In this way, the creation of a waste management program in Educational Institutions is fundamental not only as a way of minimizing the impacts caused to the environment, but mainly to educate future professionals (MICARONI, 2002).

ENVIRONMENTAL CHEMISTRY AND ENVIRONMENTAL EDUCATION

According to Baird (2002), Environmental Chemistry involves the study of a large number of different chemical agents, processes and remediation procedures. It is the study of the sources, reactions, transportation and fate of chemical species in the environments: air, water and soil, as well as the study of their effects on human health (MANO, 2010).

Environmental education is teaching that leads to the rational use of natural resources and the maintenance of an optimum level of productivity in natural ecosystems or those managed by human beings (LEITE, 2011).

It is the environmental education of being, which establishes links with the world, and that of knowing how to care, as a human ethic of compassion for the Earth and aimed at the sustainability of life on the planet. Environmental education must focus on understanding and acting on the relationships that give rise to environmental problems and not just restrict itself to informing about their consequences. The meaning of environmental education today goes beyond making people aware of the problem. It is no longer enough just to know what is right and what is wrong in relation to the environment. We need to incorporate the environmental issue into our daily actions as a priority (CUNHA; GUERRA, 2005).

Environmental education does not refer exclusively to relationships seen as natural or ecological, but rather to all the relationships that place us on the planet and that take place in society. Thus, environmental education is defined by the uniqueness of the processes that problematize the characteristics of social groups (LEITE, 2011).

According to Leme (2003), although the issue of environmental education is currently in the spotlight, especially when it comes to educational institutions, few actually put the concepts into practice, due to the insufficient training of teachers in the area of environmental chemistry.

Teacher training in environmental education must allow them to experience participatory and democratic situations. The teacher must learn by doing, by experiencing and reflecting, but above all, it is important to encourage the teacher to seek out information, to recognize where and how to access it and to know the possible channels for participation in society. This is about forming a personal and professional identity. Thus, more than training teachers in environmental education, more than training seeking to add new pedagogical skills, it challenges the formation of an ecological subject, capable of embodying the societal, ethical and aesthetic dilemmas configured by the societal crisis in its countercultural translation; tributary of a project for a socially emancipated and environmentally sustainable society. However, there is still the challenge of internalizing in the institutional spaces structuring the educational field, the formation of a sensitivity and a critical reading of environmental problems (LEME 2003; SATO; CARVALHO, 2005).

The field of Environmental Chemistry has long been warning about the damage that chemical processes can cause to the environment. The main paradigm of Green Chemistry is to reduce or eliminate the use or generation of hazardous substances during the planning, manufacture and application of chemical products (FERREIRA; ROCHA; SILVA, 2014).

Environmental chemistry expands the horizons of conventional chemistry, giving it a socio-economic dimension, as well as fostering partnerships with other areas of knowledge such as toxicology, health engineering and biology. When practiced in this way, Environmental Chemistry revives Chemistry as a natural science and should be seen as the most powerful tool in rescuing the importance of Chemistry as one of the sciences that has brought the most benefits to man (MOZETO, 2002).

Green chemistry encompasses all aspects and types of chemical processes that reduce negative impacts on human health and the environment

(MICARONI, 2002). This area of chemistry is concerned with developing technologies and processes that are incapable of causing pollution. Ideally, the application of the principles of Green Chemistry leads to regulation, control so as not to cause unnecessary remediation. In addition to the environmental benefits, this thinking also has an economic impact thanks to the reduction in expenditure on the storage and treatment of waste, decontamination and the payment of compensation. Green chemistry is a type of pollution prevention in the field of chemistry (PRADO, 2003).

According to Mano (2010), pollution is any alteration to the natural properties of the environment that is harmful to the health, safety or well-being of the population subject to its effects, caused by an agent of any kind.

The concept of pollution prevention can be defined as follows: "Any practice that reduces the quantity or environmental and health impact of any pollutant prior to its recycling, treatment or final disposal, including modification of equipment or technologies, reformulation or *redesign* of products, substitution of raw materials and organizational improvement, training or inventory control (BRAGA et al. 2005).

According to Braga (et al. 2005), the main objective of any pollution prevention initiative is to reduce the environmental impacts associated with the entire life cycle of a product or substance, with the aim of protecting our health and the environment.

Also according to Braga (et al. 2005), educational programs should be directly linked to pollution prevention, through which there is the possibility of developing awareness programs, addressing the problems associated with pollution, emphasizing the need to adopt prevention strategies, in addition to the benefits that these can provide.

WASTE MANAGEMENT

According to CONAMA Resolution No. 358 of April 29, 2005, chemical waste is any material or substance that is hazardous when not submitted to a reuse or recycling process, which may present a risk to public health or the environment, depending on its characteristics of flammability, corrosivity, reactivity and toxicity (UNIFAL, 2014).

Currently, there is worldwide concern about the environmental impacts caused by solid waste, liquid effluents and volatile compounds. This concern is the result of environmental contamination that affects public health and biodiversity and has been generated by the mismanagement of chemical waste produced (SISSINO, 2000).

In Brazil, due to the lack of specific legislation for the treatment of chemical waste, more effective control and supervision of educational, research and service-providing institutions, these generating units have done very little to minimize the production and correct disposal of potentially hazardous waste (MICARONI, 2002).

University chemistry departments and institutes, as well as all units that use chemical products in their work routines, have been confronted for many years with the problem of the treatment and final disposal of the waste generated in their teaching and research laboratories. This waste differs from that generated in industrial units in that it has a low volume but a great diversity of compositions, which makes it difficult to establish a standard and appropriate chemical treatment and/or final disposal for such chemical waste (GERBASE, 2005; GILONI-LIMA, LIMA, KATAOKA, 2011).

In addition, the inappropriate disposal of chemical waste from both practical classes and research is a problem in most universities. In addition, the amount of chemical waste generated in teaching laboratories has been increasing considerably (BENTO; PAIM, 2015; SARI et al., 2015).

The waste generated by Universities and Research Institutions causes great concern, since in Brazil around 80% of sewage does not receive any kind of treatment and is dumped directly into rivers, lakes, springs and the sea. Even if Teaching and Research Institutions are served by the public sewage treatment system, they must be very careful when disposing of their toxic waste, as conventional effluent treatments (biological and chemical sedimentation) are often effective at removing conventional pollutants, but inefficient at eliminating many toxic chemical reagents (MICARONI, 2002).

Standard NBR 10004 (ABNT, 2004) - "Solid Waste - Classification", classifies waste according to its potential risks to the environment and public health, indicating which waste should be handled and disposed of more strictly. According to this standard, waste is classified into three categories:

Class I - Hazardous waste: solid waste or mixtures of waste which, due to their characteristics of flammability, corrosiveness, reactivity, toxicity and pathogenicity, may present risks to public health, causing or contributing to adverse effects on the environment, when handled or disposed of inappropriately.

Class II - Non-inert waste: solid waste or mixtures of solid waste that may have properties such as combustibility, biodegradability or solubility in water.

Class III Waste: Inert: solid waste or mixtures of solid waste which, when subjected to solubilization tests, do not have any of their constituents solubilized in concentrations higher than the water potability standards (MACEDO, 2006).

Toxic waste is waste that can cause damage to the health of people or other living beings (BAIRD, 2002). This class includes heavy metals, which make up the majority of waste stored in educational institutions.

Metals are non-degradable substances, so they can accumulate in the components of the environment where they show their toxicity. Chemically, heavy metals are defined as a group of elements located near the bottom of the Periodic Table, more precisely between copper and lead, with high atomic masses and densities. They are highly reactive and bioaccumulative metals, meaning that the body is unable to eliminate them (BAIRD, 2002; GUERREIRO, 2013).

Chemical waste will almost always contain metal ions with toxic potential such as Copper (Cu), Iron (Fe), Manganese (Mn), Zinc (Zn), Cobalt (Co), Nickel (Ni), Vanadium (V), Aluminum (Al), Silver (Ag), Cadmium (Cd), Chromium (Cr), Mercury (Hg), Lead (Pb), etc. However, human activities have increased the levels of metal ions in many ecosystems (MACEDO, 2006).

The treatment and recovery of these metals is important because, in addition to the economic advantage, there is also the need to protect the environment. However, when investing in the treatment of waste, it is important that it is reduced at the source, because if this waste is minimized, the time and reagents used for treatment will consequently be reduced, as well as the energy spent on the process. In addition, the treatment of chemical products is not always efficient, leaving potentially toxic substances that can be disposed of directly into the environment (AHMED; NAYL; DAOUD, 2012; Sotiles et al., 2015).

Reducing the source of waste production is a preventative strategy and can be achieved with a specific policy implemented through regulatory, economic and social instruments (LIMA, 2005).

According to Maroun (2006), it is of fundamental importance to develop and implement a Waste Management Plan and, to do so, there is a need to establish some stages such as planning, implementation and operation, verification and corrective actions within this type of program. The Waste Management Plan must be planned in such a way as to define in which process the waste is produced, in which classification it falls and how much of it is generated in a given period of time. Waste management must be carried out in accordance with the 3Rs Policy - Reduce, Reuse, Recycle: reducing waste generation at source, reusing it and, finally, recycling it (APED, 2010).

Implementing the Waste Management Plan is the most difficult and time-consuming stage, as it involves everything from training the professionals involved with the waste to handling and storing it for subsequent treatment. Finally, evaluations of the planning stages must be carried out using indicators and monitoring, so that waste management is effective.

MATERIAL AND METHODS

Initially, we reformulated some of the laboratory practices in the General Chemistry textbook, which are currently being carried out in the first year of Integrated High School at a public school in a town in the southwest of Paraná. We then tested other practices from the General and Experimental Chemistry textbook from the State University of the Center West - UNICENTRO, for possible replacement or even implementation of a new experiment in the public school, which would save more on materials and reagents.

After determining the practices to be used, they were reviewed in order to ensure that the smallest possible quantities of reagents were used, seeking the best cost-benefit ratio. These experiments aimed to reduce the cost of reagents, reduce the volume of waste generated and consequently reduce the cost of treating this waste, as well as reducing the environmental impact caused by incorrect waste disposal.

EXPERIMENT 1: CHEMICAL REACTIVITY OF METALS

When a zinc strip (Zn) is placed in a solution of copper sulphate (Cuso4), the Zn becomes covered in a dark red substance (metallic copper), while the Zn^2 ions+ migrate into the solution, which becomes colorless.

The reaction occurs because Zn is more reactive than copper (Cu), i.e. it has a greater capacity to donate electrons. Table 1 shows the chemical reactivity series of metals.

Table 1 - Metal chemical activity series

Reduced state	Oxidized state	F.E.M. (Volts)
Li	Li^+	+3,02
Rb	Rb^+	+2,99
K	K^+	+2,92
Ba	Ba^{2+}	+2,90
mr	Sr^{2+}	+2,89
Ca	Ca^{2+}	+2,87
In	In^+	+2,71
Mg	Mg^{2+}	+2,34
Al	Al^{3+}	+1,67
Mn	Mn^{2+}	+1,05
Zn	Zn^{2+}	+0,76
Cr	Cr^{2+}	+0,71
Fe	Fe^{2+}	+0,44
Co	Co^{2+}	+0,28
Ni	Ni^{2+}	+0,25
sn	Sn^{2+}	+0,14
Pb	Pb^{2+}	+0,13
H	H^+	0,00
Bi	Bi^{3+}	-0,32
Cu	Cu^+	-0,35
Hg	Hg^{2+}	-0,80
Ag	Ag^+	-0,80
Hg	Hg^{2+}	-0,85
Pt	Pt^{2+}	-1,20
Au	Au^{3+}	-1,42

In order for a reaction involving electron transfer to take place, it is necessary to take into account the chemical reactivity series of metals, since they tend to lose electrons, i.e. oxidize, and the one that is easiest to donate electrons is the most reactive.

The aim of the practical is to observe and compare the reactivity and behavior of metals towards acidic and basic substances and salts.

MATERIALS AND REAGENTS

- Test tubes
- Copper sulphate ($CuSO_4$) 1 mol L^{-1}
- Zinc sheets (Zn)
- Sodium chloride (NaCl) 1 mol L^{-1}
- Hydrochloric acid (HCl) 0.5 mol L^{-1}
- Silver nitrate ($AgNO_3$) 1 mol L^{-1}
- Sodium hydroxide (NaOH) 1 mol L^{-1}
- Concentrated nitric acid (HNO_3)
- Aluminum (Al), Zn, Cu, iron (Fe) and magnesium (Mg) shavings

EXPERIMENTAL PROCEDURE

Reactions of metals with salts

Place a zinc slide in three test tubes.

To the first, add 2 mL of $CuSO_4$ solution.

Add 2 mL of NaCl solution to the second.

Add 2 mL of $AgNO_3$ solution to the third.

Wait 10 minutes, shake and observe.

Reactions of metals with acids

Number five test tubes and place 2 mL of 0.5 mol L^{-1} HCl in each tube. Add Al, Zn, Cu, Fe and Mg chips, respectively, to each tube.

Sort the metals in increasing order of reactivity according to the time taken for the reaction.

Place a small piece of Cu in a test tube and add 2 mL of concentrated HNO_3.

Reaction of metals with hydroxides

Number 3 test tubes and put 3 mL of sodium hydroxide (NaOH) in each one.

Add a small piece of Al to the first, Cu to the second and Fe to the third.

Warm up and observe.

RESULTS

Metal reactions with salts

After adding zinc and 0.2 mL of $CuSO_4$, NaCl and $AgNO_3$ solutions (Figure 1), it was observed that only NaCl did not react, since it is more reactive than Zn. In the case of $CuSO_4$ (Equation 1) and $AgNO_3$ (Equation 2), the ions in the solution adhered to the plate, while those in the Zn migrated into the solution.

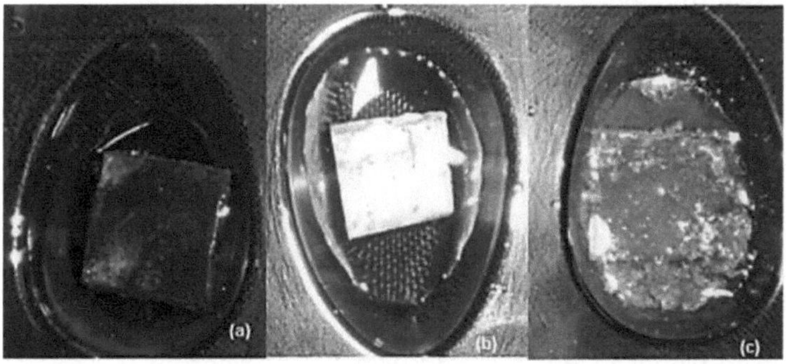

Figure 1 - Reaction of zinc with salts (a)$CuSO_4$, (b)NaCl, (c)$AgNO_3$.

$Zn_{(s)} + CuSO_{4(aq)} \rightarrow ZnSO_{4(aq)} + CU_{(S)}$ (Equation 1)

$Zn_{(s)} + 2\ AgNO_{3(aq)} \rightarrow 2Ag_{(s)} + Zn(NO_3)_{2(s)}$ (Equation 2)

Reactions of metals with acids

With the addition of 0.2 mL of HCl, it was possible to observe the reactivity of the metals (Figure 2). Some did not occur instantaneously, but after a few minutes, the reaction began.

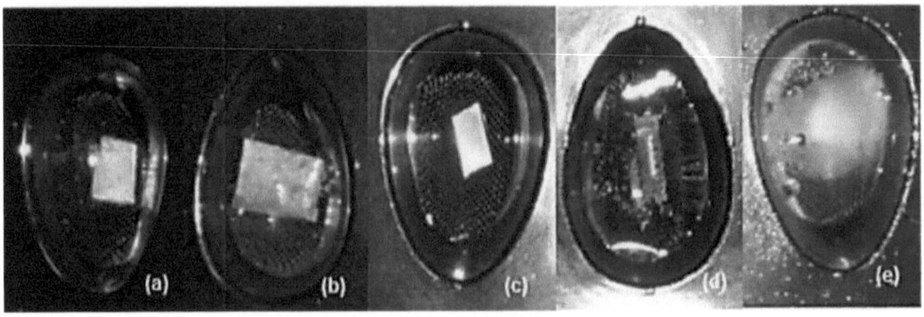

Figure 1 - Reactivity of metals with HCl. (a) Al, (b) Zn, (c) Cu, (d) Fe, (e) Mg.

After carrying out the experiment, it was found that the order of reactivity of the metals in ascending order is:

Mg Zn> > Al> Fe> Cu

The reactions of metals with acids can be found in Equations 3 to 6.

$$Mg_{(s)} + 2HCl_{(aq)} \rightarrow MgCl_{2(aq)} + H_{2(g)} \quad \text{(Equação 3)}$$

$$Zn_{(s)} + 2HCl_{(aq)} \rightarrow ZnCl_{2(aq)} + H_{2(g)} \quad \text{(Equação 4)}$$

$$2Al_{(s)} + 6HCl(aq) \rightarrow 2AlCl_{3(aq)} + 3H_{2(g)} \quad \text{(Equação 5)}$$

$$Fe_{(s)} + 2HCl_{(aq)} \rightarrow FeCl_{2(aq)} + H_{2(g)} \quad \text{(Equação 6)}$$

Reaction of metals with hydroxides

It was also possible to visualize the reaction with the addition of 0.2 mL of NaOH.

Figure 2 - Reactivity of metals with hydroxide: (a) Al, (b) Cu, (c) Fe

The reaction with Al occurs spontaneously, according to the reaction described in Equation 7.

$$Al_{(s)} + NaOH + H_2O \rightarrow NaAlO_2 + 3/2\ H_{2(g)} \qquad \text{(Equação 7)}$$

In the case of Cu and Fe, however, the reaction doesn't occur even with heating, as sodium is more reactive and the exchange of ions doesn't take place.

EXPERIMENT 2: CHEMICAL REACTIONS

When one or more substances change into another, we call it a chemical reaction.

A chemical equation is the representation of a reaction, showing the reactants (substances that react) and the products (substances formed).

Reagents ⟶ ***Products***

Synthesis, formation or addition reaction

These are chemical reactions in which two or more substances react to form a single substance.

The synthesis reaction can be:

Total: when all the reactants are simple substances.

Partial: when there are one or more compound substances.

Analysis or decomposition reaction

These are chemical reactions in which a single reactant forms two or more substances.

$$A + B \longrightarrow AB$$

Displacement or Simple Exchange Reaction

These are chemical reactions in which a simple substance reacts with a compound substance by displacing one of its elements, giving rise to another simple and compound substance.

$$AB \longrightarrow A + B$$

Double Exchange Reaction

These are chemical reactions in which compound substances exchange their cations or anions to form two other compound substances.

$$C + AB \longrightarrow B + AC$$

The aim of this practical is to construct/write the chemical equation for each reaction and identify them according to their classification: synthesis, decomposition, simple or double exchange reactions.

MATERIALS AND REAGENTS

- Test tubes
- 10 mL pipettes
- Tube holder
- Washer bottle
- Tube brush
- Spatula
- Matchsticks
- NaCl solution
- Phenolphthalein
- Wooden tweezers
- HCl solution 1 mol L^{-1}

- Silver nitrate solution ($AgNO_3$) 1 mol L^{-1}
- Copper sulphate solution ($CuSO_4$) 1 mol L^{-1}
- Chlorinated water
- Magnesium Blade (Mg)
- Cu foil
- Zn blade
- Al foil
- Potassium iodide solution (KI) 1 mol L^{-1}
- NaOH solution 1 mol L^{-1}
- Sodium thiosulfate solution ($Na_2S_2O_3$) 1 mol L^{-1}
- Ammonium hydroxide solution (NH_4OH) 1 mol L^{-1}

EXPERIMENTAL PROCEDURE

a) Mg reacts with oxygen (O) to form an oxide. To observe this reaction, proceed as follows: place a piece of Mg on the tip of a spatula and heat it with a Bunsen burner until a white powder, Mg oxide (MgO), appears.

b) MgO is an oxide which, in contact with water, forms a base. To observe the reaction: place the oxide obtained in item a in a test tube, add 10 mL of deionized water and a drop of phenolphthalein indicator. The appearance of a pink color proves that the medium is basic.

c) Place 0.5 mL of $AgNO_3$ solution and a Cu slide in a test tube. Wait a few minutes and observe what happens.

d) Place 2 mL of CuSO4 and a previously cleaned Zn slide in another test tube. Wait a few minutes and observe what happens.

e) In a third test tube, pour 2 mL of NaCl solution and add a slide of Al. Wait and observe what happens.

f) Place 2 mL of KI solution in a fourth test tube, add 1 mL of chlorinated water and observe what happens.

g) Place 5 mL of NH4OH solution in a test tube and bring it to the flame of the Bunsen burner and smell the characteristic odor of ammonia gas.

h) Place 2 mL of NaOH solution and two drops of phenolphthalein indicator in 1 test tube. Add 2 mL of HCl solution drop by drop and observe when the color of the indicator disappears. Analyze the reaction.

i) Place 2 mL of CuSO4 solution in a test tube and add 2 mL of NaOH solution. Observe the formation of a blue precipitate of cupric hydroxide. Heat the test tube until a black precipitate forms. Analyze the reactions.

j) Place 2 mL of Na2S2O3 solution and a drop of phenolphthalein indicator in a test tube. Add 2 mL of HCl and observe the action of the indicator. Analyze the reaction.

k) Leave the test tube in the holder and wait approximately 2 minutes. Observe the appearance of a whitish color and a characteristic smell of sulfur. Analyze the reaction.

RESULTS

In the reaction of metallic copper with 0.2 mL of $AgNO_3$, a simple exchange reaction is observed, in which the silver adheres to the metal while the solution turns blue due to the formation of copper nitrate ($Cu(NO_{3)(2)})$, according to Equation 8 and Figure 4.

$$2AgNO_{3(aq)} + Cu_{(s)} \rightarrow 2Ag_{(s)} + Cu(NO_3)_{2(aq)} \qquad \text{(Equação 8)}$$

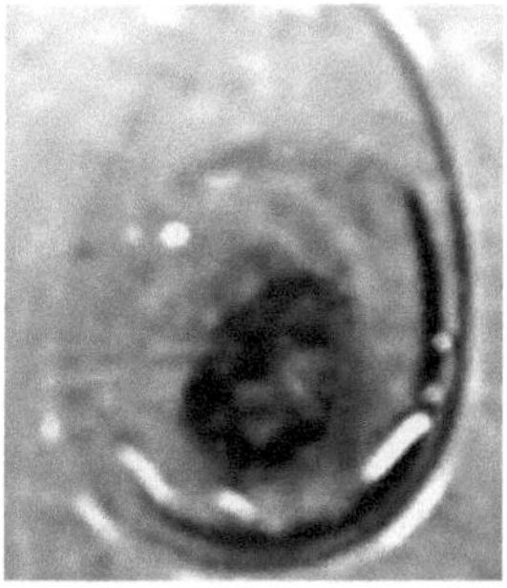

Figure 4 - Formation of copper nitrate

CuSO4 is a salt which, when in an alkaline medium, forms a gelatinous precipitate of the metallic ion in the form of cupric hydroxide ($Cu(OH)_2$), according to Equation 9 and Figure 5.

$$CuSO_{4(aq)} + 2\ NaOH_{(aq)} \rightarrow Cu(OH)_{2(s)} + Na_2SO_{4(aq} \qquad \text{(Equação 9)}$$

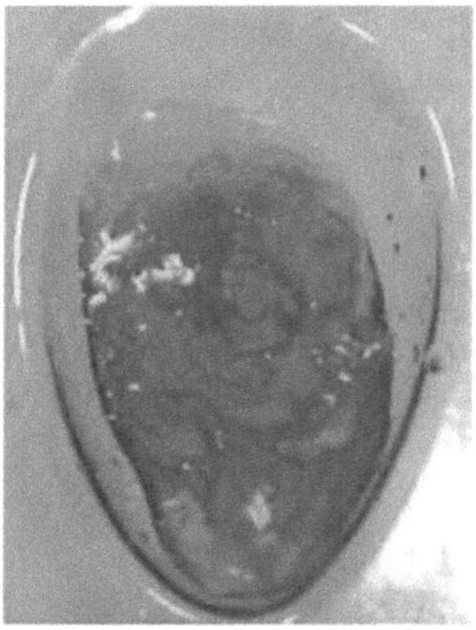

Figure 3 - Formation of cupric hydroxide

EXPERIMENT 3: PRECIPITATION REACTIONS I

Reactions that result in the formation of a poorly soluble or insoluble compound are known as precipitation reactions. Precipitation reactions occur when certain pairs of oppositely charged ions attract each other so strongly that they form an ionic solid.

The aim of this practical is to carry out a series of reactions and observe in which reactions insoluble compounds are formed.

MATERIALS AND REAGENTS

- Beakers
- Pipettes and pears
- $CuSO_4$ solution 1 mol L^{-1}
- 1 mol NaCl solution L^{-1}
- KI solution 1 mol L^{-1}
- $AgNO_3$ solution 1 mol L^{-1}
- NaOH solution 1 mol L^{-1}
- Solutionsolution chloride solution ($CaCl_2$) 1 mol L^{-1}
- Solutionsolution zinc solution ($ZnSO_4$) 1 mol L^{-1}
- Solution ofbarium solution ($BaCl_2$) 1 mol L^{-1}
- Solutionnitratelead solution ($Pb(NO_3)_{(2)}$) 1 mol L^{-1}
- KI ofpotassium solution (K_2SO_4) 1 mol L^{-1}
- Sodium carbonate solution (Na_2CO_3) 1 mol L^{-1}
- Potassium chromate solution (K_2CrO_4) 1 mol L^{-1}

EXPERIMENTAL PROCEDURE

Place 10 mL of the CuSO4 solution in five containers. Add 10 mL of solution to each container, adding NaCl to the first, Na_2CO_3 to the second, NaOH to the third, K_2SO_4 to the fourth and KI to the last. Observe what happens and work out each of the reactions. Repeat the procedure for the ZnSO4, AgNO3, CaCl2 and BaCl2 solutions.

RESULTS

After adding 0.2 mL of the solutions according to the procedure, you can see that the precipitate has formed in some, while in others, nothing has happened.

The reactions that took place with CuSO4 are shown in Figure 6 and in

Equations 10 to 14.

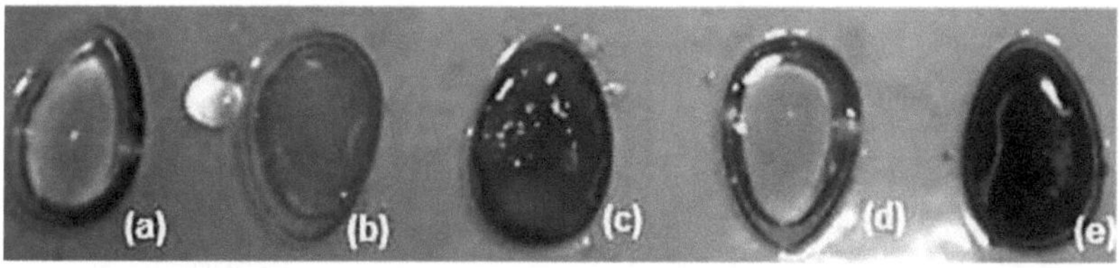

Figure 4 - Reactions of $CuSO_4$ with: (a) NaCl, (b) Na_2CO_3, (c) NaOH, (d) K_2SO_4, (e) KI.

$$CuSO_{4(aq)} + 2\ NaCl_{(aq} \rightarrow Na_2SO_{4(aq)} + CuCl_{2(aq)} \quad \text{(Equação 10)}$$

$$CuSO_{4(aq)} + Na_2CO_{3(aq)} \rightarrow CuCO_{3(s)} + Na_2SO_{4(aq)} \quad \text{(Equação 11)}$$

$$CuSO_{4(aq)} + 2\ NaOH_{(aq)} \rightarrow Na_2SO_{4(aq)} + Cu(OH)_{2(s)} \quad \text{(Equação 12)}$$

$$CuSO_{4(aq)} + K_2SO_{4(aq)} \rightarrow CuSO_{4(aq)} + K_2SO_{4(aq)} \quad \text{(Equação 13)}$$

$$CuSO_{4(aq)} + 2\ KI_{(aq)} \rightarrow K_2SO_{4(aq)} + CuI_{2(aq)} \quad \text{(Equação 14)}$$

For $ZnSO_4$, the behavior was similar to that of $CuSO_4$, and it was possible to observe the formation of the precites in the form of carbonate and hydroxide, as shown in Figure 7 and in Equations 15 to 19.

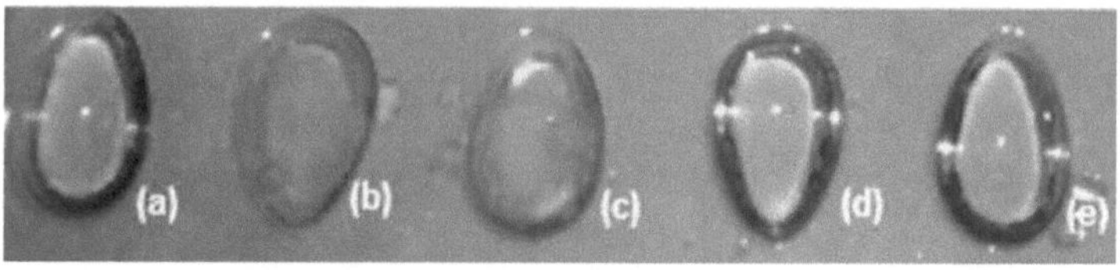

Figure 5 - Reactions of ZnSO4 with: (a) NaCl, (b) Na_2CO_3, (c) NaOH, (d) K_2SO_4, (e) KI.

$$ZnSO_{4(aq)} + 2\ NaCl_{(aq} \rightarrow Na_2SO_{4(aq)} + ZnCl_{2(aq)} \quad \text{(Equação 15)}$$

$$ZnSO_{4(aq)} + Na_2CO_{3(aq)} \rightarrow ZnCO_{3(s)} + Na_2SO_{4(aq)} \quad \text{(Equação 16)}$$

$$ZnSO_{4(aq)} + 2\ NaOH_{(aq)} \rightarrow Na_2SO_{4(aq)} + Zn(OH)_{2(s)} \quad \text{(Equação 17)}$$

$$ZnSO_{4(aq)} + K_2SO_{4(aq)} \rightarrow ZnSO_{4(aq)} + K_2SO_{4(aq)} \quad \text{(Equação 18)}$$

$$ZnSO_{4(aq)} + 2\ KI_{(aq)} \rightarrow K_2SO_{4(aq)} + ZnI_{2(aq)} \quad \text{(Equação 19)}$$

In the case of silver, precipitates formed with all the solutions, and the reactions are shown in Equations 20 - 24 and Figure 8.

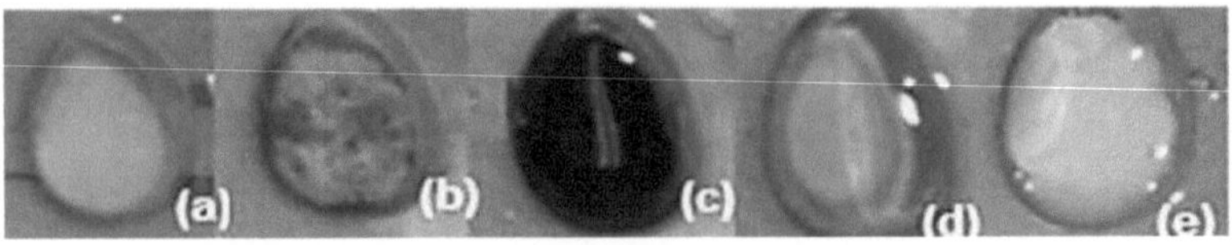

Figure 6 - Reactions of AgNO3 with: (a) NaCl, (b) Na2CO3, (c) NaOH, (d) K2SO4, (e) KI.

$$AgNO_{3(aq)} + NaCl_{(aq)} \rightarrow AgCl_{(s)} + NaNO_{3(aq)} \quad \text{(Equação 20)}$$

$$AgNO_{3(aq)} + Na_2CO_{3(aq)} \rightarrow Ag_2CO_{3(s)} + 2\ NaNO_3 \quad \text{(Equação 21)}$$

$$AgNO_{3(aq)} + NaOH_{(aq)} \rightarrow AgOH_{(s)} + NaNO_{3(aq)} \quad \text{(Equação 22)}$$

$$2\ AgNO_{3(aq)} + K_2SO_{4(aq)} \rightarrow Ag_2SO_{4(aq)} + 2\ KNO_{3(aq)} \quad \text{(Equação 23)}$$

$$AgNO_{3(aq)} + KI \rightarrow AgI_{(s)} + KNO_{3(aq)} \quad \text{(Equação 24)}$$

Calcium chloride and barium chloride showed similar behavior to the results obtained in this experiment (Figures 9 and 10).

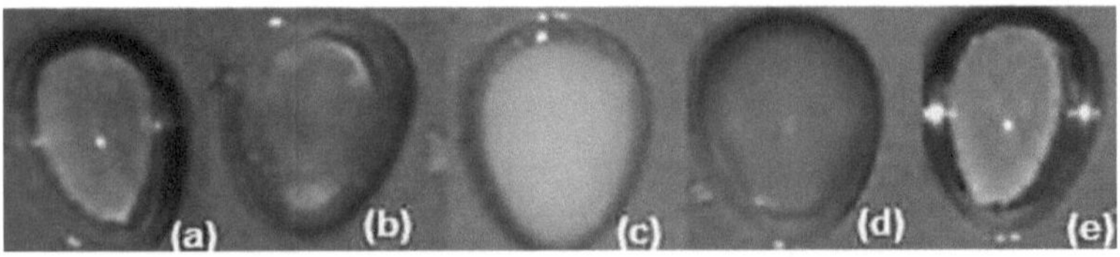

Figure 7 - Reactions of CaCl2 with: (a) NaCl, (b) Na2CO3, (c) NaOH, (d) K2SO4, (e) KI.

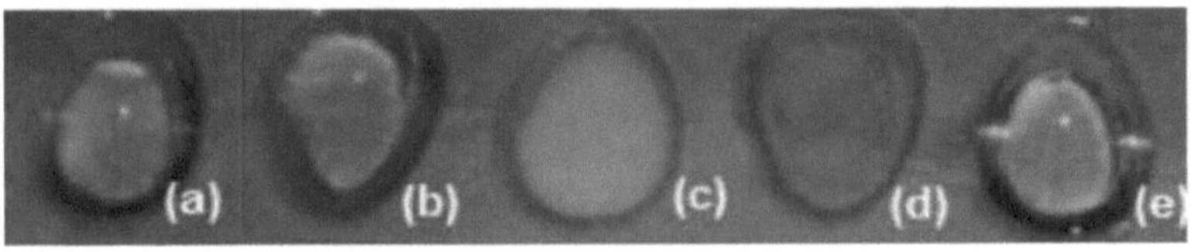

Figure 8 - Reactions of BaC | 2 with: a) NaCl, (b) Na2CO3, (c) NaOH, (d) K2SO4, (e) KI

The reactions of calcium and barium with the other substances are shown in Equations 25 to 29 and Equations 30 to 34, respectively.

$$CaCl_{2(aq)} + NaCl_{(aq)} \rightarrow CaCl_{2(aq)} + NaCl_{(aq)} \quad \text{(Equação 25)}$$

$$CaCl_{2(aq)} + Na_2CO_{3(aq)} \rightarrow CaCO_{3(s)} + 2\ NaCl_{(aq)} \quad \text{(Equação 26)}$$

$$CaCl_{2(aq)} + 2\ NaOH_{(aq)} \rightarrow Ca(OH)_{2(s)} + 2\ NaCl_{(aq)} \quad \text{(Equação 27)}$$

$$CaCl_{2(aq)} + K_2SO_{4(aq)} \rightarrow CaSO_{4(s)} + 2\ KCl_{(aq)} \quad \text{(Equação 28)}$$

$$CaCl_{2(aq)} + 2\ KI_{(aq)} \rightarrow CaI_{2(aq)} + 2\ KCl_{(aq)} \quad \text{(Equação 29)}$$

$$BaCl_{2(aq)} + NaCl_{(aq)} \rightarrow BaCl_{2(aq)} + NaCl_{(aq)} \quad \text{(Equação 30)}$$

$$BaCl_{2(aq)} + Na_2CO_{3(aq)} \rightarrow BaCO_{3(s)} + 2\ NaCl_{(aq)} \quad \text{(Equação 31)}$$

$$BaCl_{2(aq)} + 2\ NaOH_{(aq)} \rightarrow Ba(OH)_{2(s)} + 2\ NaCl_{(aq)} \quad \text{(Equação 32)}$$

$$BaCl_{2(aq)} + K_2SO_{4(aq)} \rightarrow BaSO_{4(s)} + 2\ KCl_{(aq)} \quad \text{(Equação 33)}$$

$$BaCl_{2(aq)} + 2\ KI_{(aq)} \rightarrow BaI_{2(aq)} + 2\ KCl_{(aq)} \quad \text{(Equação 34)}$$

EXPERIMENT 4: PRECIPITATION REACTIONS II

Precipitation reactions have many applications. One of them is the production of compounds. The strategy is to choose solutions that provide, when mixed, a precipitate of the desired insoluble compound, separating it by filtration.

In qualitative analysis, the formation of a precipitate is used to confirm the identity of certain ions. In quantitative analysis, the aim is to determine the amount of each substance or element present in the sample. Quantitative analysis is widely used in environmental monitoring to find out how much lead or mercury is in a water sample.

The aim of this practical is to observe a precipitation reaction and, through the experiments, discuss the solubility of ionic solids and compare it with their periodic properties.

MATERIALS AND REAGENTS

- Test tubes -Chloridechloride calcium chloride ($CaCl_2$) 1 mol L^{-1}
- 100 mL beaker. -Chloridechloride strontium chloride ($SrCl_2$) 1 mol L^{-1}
- Distilled water - Barium chloride ($BaCl_2$) 1 mol L^{-1}
- Pipettes and pears -Chloridechloride aluminum chloride ($AlCl_3$) 1 mol L^{-1}
- Glass bastâo -Chloridechloridechloride ($MnCl_2$) 1 mol L^{-1}
- Sodiumsodium chloride (NaCl) 1 mol L^{-1} -Sodiumzinc zinc chloride ($ZnCl_2$) 1 mol L^{-1}
- Potassium potassium chloride (KCl) 1 mol L^{-1} - Chloride silver silver chloride (AgCl) 1 mol L^{-1}
- Magnesiummagnesium chloride ($MgCl_2$) 1 mol L^{-1} - Chloride lead lead II chloride ($PbCl_2$) 1 mol L^{-1}
- Ammoniumammonium chloride (NH_4Cl) 1 mol L^{-1}- Chloride Lithium Lithium chloride (LiCl) 1 mol L^{-1}

- Solutions of fluoride, bromide, iodide, nitrate, carbonate, sulphate, phosphate, chromate and hydroxide, all at a concentration of 1 mol L^{-1}.

EXPERIMENTAL PROCEDURE

Place 10 mL of the NaCl solution in different beakers. Add 10 mL of the fluoride, bromide, iodide, nitrate, carbonate, sulphate, phosphate, chromate and hydroxide solutions, one in each container. Repeat the procedure for the solutions of KCl, LiCl, NH_4Cl, AgCl, $MgCl_2$, $PbCl_{(2)}$, $ZnCl_{(2)}$, $BaCl_2$, $MnCl_{(2)}$, $AlCl_3$, $SrCl_{(2)}$, $CaCl_{(2)}$. Observe what happens and work out the reactions.

RESULTS

Just like the reactions in the previous precipitation practice, the results could also be visualized with just 0.2 mL of each reagent.

For lithium, sodium, potassium and ammonium chlorides, the reactions were the same, according to Figure 11 and Equations 35 to 43, in which the anions are represented by an X.

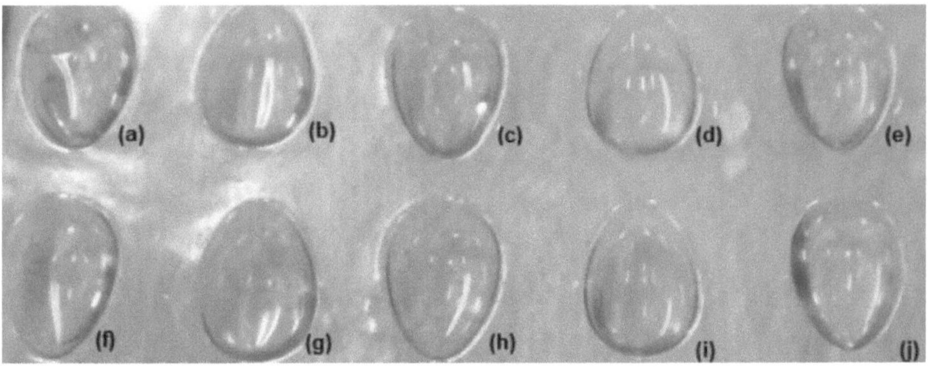

Figure 9 - Reactions of sodium, lithium, potassium and ammonium chlorides with: (a) F^-, (b) $Cl^{(-)}$,
(c) Br^-, (d) I^-, (e) NO_3^-, (f) SO_4^{-2}, (g) CO_3^{-2}, (h) $PO_{(4)}^{(-3)}$), (i) $CrO_{(4)}^{)(-2)}$, (j) $OH_{(^-)}$)

$XCl + F^{-} \rightarrow XF + Cl^{-}$ (Equação 35)

$XCl + Br^{-} \rightarrow XBr + Cl^{-}$ (Equação 36)

$XCl + I^{-} \rightarrow XI + Cl^{-}$ (Equação 37)

$XCl + NO_3^{-} \rightarrow XNO_3 + Cl^{-}$ (Equação 38)

$2\ XCl + SO_4^{-2} \rightarrow X_2SO_4 + 2\ Cl^{-}$ (Equação 39)

$2\ XCl + CO_3^{-2} \rightarrow X_2CO_3 + 2\ Cl^{-}$ (Equação 40)

$3\ XCl + PO_4^{-3} \rightarrow X_3PO_4 + 3\ Cl^{-}$ (Equação 41)

$2\ XCl + Cr_2O_4^{-2} \rightarrow X_2Cr_2O_4 + 2\ Cl^{-}$ (Equação 42)

$XCl + OH^{-} \rightarrow XOH + Cl^{-}$ (Equação 43)

The chlorides of magnesium, calcium, strontium, barium, aluminum, zinc, silver and manganese also showed the same behavior, according to the reactions described in Equations 44 to 52 in which the anions are represented by a Y and Figure 12.

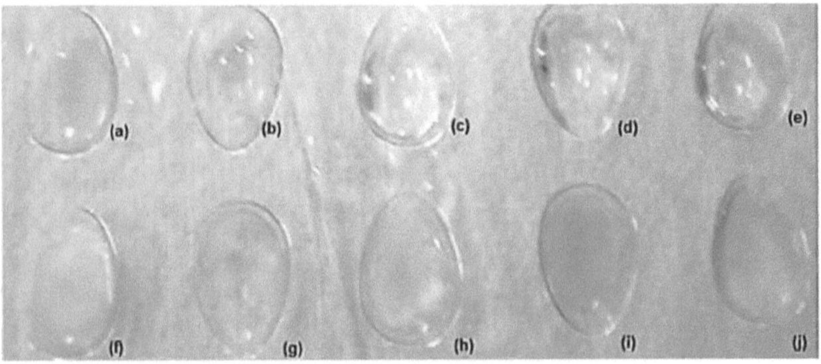

Figure 10 - Reactions of the other chlorides with (a) $F^{(-)}$, (b) $Cl^{(-)}$), (c) $Br^{(-)}$), (d) $I^{(-)}$), (e) $NO_{(3)(}^{-}$), (f) SO_4^{-2}, (g) $CO_{(3)(}^{-2})$, (h) $PO_{(4}^{)(-3}$), (i) $CrO_{(4}^{)(-2)}$, (j) $OH_{(}^{-}$)

$YCl_2 + 2\ F^{-} \rightarrow YF_2 + 2\ Cl^{-}$ (Equação 44)

$Y\ Cl_2 + 2\ Br^{-} \rightarrow YBr_2 + 2\ Cl^{-}$ (Equação 45)

$YCl_2 + 2\ I^{-} \rightarrow YI_2 + 2Cl^{-}$ (Equação 46)

$YCl_2 + NO_3^{-} \rightarrow Y(NO_3)_2 + 2\ Cl^{-}$ (Equação 47)

$YCl_2 + SO_4^{-2} \rightarrow YSO_4 + 2\ Cl^-$ (Equação 48)

$YCl_2 + CO_3^{-2} \rightarrow YCO_3 + 2\ Cl^-$ (Equação 49)

$YCl_2 + PO_4^{-3} \rightarrow YPO_4 + 2\ Cl^-$ (Equação 50)

$YCl_2 + Cr_2O_4^{-2} \rightarrow YCr_2O_4 + 2\ Cl^-$ (Equação 51)

$YCl_2 + 2\ OH^- \rightarrow Y(OH)_2 + 2\ Cl^-$ (Equação 52)

EXPERIMENT 5: CHEMICAL KINETICS - LANDOLT REACTION

There is a branch of science that studies the speed of chemical reactions and the factors that influence them, called Chemical Kinetics. Chemical reactions can be defined as a set of phenomena in which two or more substances react with each other, giving rise to different compounds.

The speed of a reaction is how quickly the reactants are consumed or how quickly the products are formed. The combustion of a candle and the formation of rust are examples of slow reactions. In dynamite, the decomposition of nitroglycerin is a fast reaction.

The speeds of reactions are determined through empirical laws, called rate laws, deduced from the effect of the concentration of reactants and products on the speed of the reaction.

$v= k.[A]^a[B]^b$

Where [A] and [B] represent the concentrations of the reactants, k is a temperature-dependent constant and a and b correspond to the order of reaction for each reactant, and can only be determined experimentally.

The aim of this practical is to observe the different reaction speeds depending on the concentration of the reactants.

MATERIALS AND REAGENTS

- Beakers and beakers
- Starch 4 g L^{-1}
- Sulfuric acid at 1 mol L^{-1}
- Solution A: Potassium iodate (KIO_3) 4.3 g L^{-1}
- Solution B: Sodium bisulfite ($NaHSO_3$) 1 g $L^{(-1)}$

EXPERIMENTAL PROCEDURE

Number 10 beakers. Add 10 mL of solution A to beaker 1 and then to the others, following the volumes in Table 2. Add the amount of water as indicated in the table. Then add solution B to the respective beakers and set the time for the reaction to take place.

Table 2 - Quantity of each solution

Solution	Solution A (mL)	H_2O (mL)	Solution B (mL)
1	10,0	-	5,0
2	9,0	1,0	5,0
3	8,0	2,0	5,0
4	7,0	3,0	5,0
5	6,0	4,0	5,0
6	5,0	5,0	5,0
7	4,0	6,0	5,0
8	3,0	7,0	5,0
9	2,0	8,0	5,0
10	1,0	9,0	5,0

Post-lab - Discussions

a) Calculate the concentration of KIO3 in each tube, writing the values down in a table and sketch a graph of the concentration versus reaction time and define the reaction order and the reaction rate constant.

b) Write down the reactions that took place in the experiment and analyze the results based on kinetic theory.

RESULTS

After the reaction that takes place between solutions A (potassium iodate) and B (sulfuric acid, sodium bisulfite and starch indicator) in the proportions indicated in Table 2, the time needed to visualize the occurrence of the reaction was determined (Figure 13).

Figure 11 - Reaction between the reagents potassium iodate, sodium bisulphite and sulphuric acid in Landolt's kinetics experiment

The time readings are shown in Table 3.

Table 3- Results of the time required for the reaction to occur

Tube	Solution A (mL)	H_2O (mL)	Solution B (mL)	Marked time (s)
1	1,0	-	2,0	1,00
2	0,9	0,1	2,0	1,38
3	0,8	0,2	2,0	1,62
4	0,7	0,3	2,0	2,18
5	0,6	0,4	2,0	2,69
6	0,5	0,5	2,0	3,45
7	0,4	0,6	2,0	4,13
8	0,3	0,7	2,0	5,20
9	0,2	0,8	2,0	6,95

10	0,1	0,9	2,0	Not detected

The concentration of KIO_3 in each tube was determined, as well as the number of moles of reactant over time (Table 4).

Table 4 - KIO_3 results

	Concentration of solution A (mol L^{-1})	**Number of moles of KIO_3 that reacted**	**Reaction time (s)**	**Reaction speed (mol s^{-1})**
1	2,01 E-05	6,028 E-05	1,00	6,028 E-05
2	1,81 E-05	5,425 E-05	1,38	3,931 E-05
3	1,61 E-05	4,822 E-05	1,62	2,977 E-05
4	1,41 E-05	4,219 E-05	2,18	1,935 E -05
5	1,21 E-05	3,616 E-05	2,69	1,344 E-05
6	1,00 E-05	3,014 E-05	3,45	8,736 E-06
7	8,04 E-06	2,411 E-05	4,13	5,838 E-06
8	6,03 E-06	1,808 E-05	5,20	3,477 E-06
9	4,02 E-06	1,205 E-05	6,95	1,735 E-06
10	2,01 E-06	6,028 E-06	-	-

The Landolt kinetics reaction is described in Equation 53, where the formation of I_2 leads to the yellowish color acquired at the end of the experiment.

$$2\,IO_3^-{}_{(aq)} + HSO_3^-{}_{(aq)} + 9\,H^+{}_{(aq)} \rightarrow I_{2(l)} + SO_4^{-2}{}_{(aq)} + 5\,H_2O_{(l)} \qquad \text{(Equação 53)}$$

WASTE GENERATION

After carrying out the experiments, the total volume of waste normally generated and the volume to be generated with the adoption of the new practices were quantified. The figures are shown in Table 5.

Table 5- Volumes of waste generated before and after the reformulation of practical classes

Waste	**Initial volume used (mL)**	**Volume after readjustment (mL)**
$CuSO_4$	112	11,2
$AgNO_3$	96	9,6
Chrome	16	1,6
Lead	32	3,2
Bàrio	32	3,2
Strontium	16	1,6
Manganese	16	1,6
Total	320	32,0

Through the tests carried out, it was possible to reduce the volume of reagents used and, consequently, reduce the volume of waste generated to 10% of the amount initially used, as shown in Figure 14.

After carrying out the adapted practices, it can be seen that they can be carried out using a small amount of material and can be visualized even on small scales and with reactions involving heavy metals.

For the treatment of chemical waste, in addition to the expenditure of time and energy, other materials are used, leading to costs in the proper disposal of this waste.

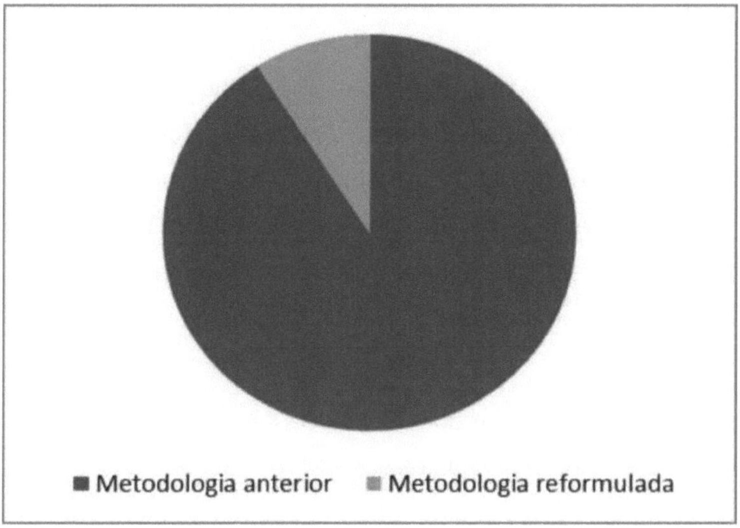

Figure 12 - Comparison of waste generated in the previous and reformulated methodologies

Reducing the amount of reagents used in chemistry experiments not only helps to reduce the environmental impact caused by the formation of toxic substances, it also helps to reduce laboratory costs and helps to raise students' ecological awareness.

CONCLUSION

Reducing the source of waste is of fundamental importance in educational institutions, as the waste treatment system can often be inefficient, failing to remove all the toxic substances present and resulting in inappropriate disposal into sewers and the public sewage system. This reduction must be based on the initial awareness of teachers and their students, based on the principles of Environmental Chemistry linked to those of Green Chemistry, with a view to Environmental Education.

After carrying out this work, it was possible to reduce the volume of reagents used by 90%, without impairing the visualization of the reactions and without compromising the students' understanding, and consequently, there was the same percentage reduction in the residues formed.

Due to the lack of specific legislation for the treatment of chemical waste and the lack of supervision of teaching and research institutions, very little has been done to minimize the generation and proper disposal of hazardous waste.

Especially when it comes to educational institutions, there is a need for educational programs guided by Environmental Chemistry, aimed at preventing pollution and reducing environmental impacts, in order to protect our health and the environment.

REFERENCES

ABNT. Brazilian Association of Technical Standards. NBR 10004. **Solid Waste - Classification.** Rio de Janeiro, 2004.

AHMED, I. M.; NAYL, A. A. & DAOUD, J. A. Leaching and recovery of zinc and copper from brass slag by sulfuric acid. **Journal of Saudi Chemical Society,** In press. 1-6, 2012

APED - Portuguese Association of Distribution Companies. **Guide to Good Environmental Practices.** Lisbon, 2010.

ATKINS, P.; JONES, L. **Principles of Chemistry: questioning modern life and the environment**. Trad. By Ricardo Bicca de Alencastro. 5ª Edition. Porto Alegre: Bookman, 2012.

BAIRD, C, **Environmental Chemistry**. Translated by Maria Angeles Lobo Recio and Luis Carlos Marques Carrera. 2nd Edition. Porto Alegre: Bookman, 2002.

BENTO, W. A. S. & PAIM, A. P. S. Treatment of copper, silver, lead, chromium and potassium permanganate waste generated in a chemistry teaching laboratory at UFPE. **Ambiciência**, v. 11, n. 1, p. 237-246, 2015.

BRAGA, B. et al. **Introduction to Environmental Engineering: The challenge of sustainable development.** Sâo Paulo: Pearson Prentice Hall, 2005.

CONAMA - NATIONAL ENVIRONMENTAL COUNCIL. Treatment and final disposal of health service waste. Resolution No. 358 of April 29, 2005.

CUNHA, S. B.; GUERRA, A. J. T. **A questao ambiental: diferentes abordagens.** Rio de Janeiro: Bertrand Brasil, 2005.

FARIAS, C.S.; BASAGLIA, A. M.; ZIMMERMANN, A. A **importância das atividades experimentais no Ensino de Quimica.** 1st Paraná Congress of Education in Chemistry, 2009.

FERRAZ, F.C.; FEITOSA, A.C. **Safety Techniques in Laboratories: Rules and Practices.** Sâo Paulo: Hemus, 2003.

FERREIRA, V. F.; ROCHA, D. R. & SILVA, F. C. Green Chemistry, Sustainable Economy and Quality of Life. **Revista Virtual de Quimica**, v. 6, n. 1; p. 85-111, 2014.

GERBASE, A. E.; COELHO, F. S.; MACHADO, P. F. L.; FERREIRA, V. F. Management of Chemical Waste in Teaching and Research Institutions. **Quimica Nova**, v. 28, n. 1, 3, 2005.

GILONI-LIMA, P. C.; LIMA, V. A. de; KATAOKA, A. M. Overview management chemical residues of laboratories in academic institutions in Brazil. In: BRONIEWICZ, E. **Environmental management in practice**. Croatia: InTech, 2011, p. 351-370.

GIORDAN, M. O papel da experimentação no ensino de ciências. **Quimica nova na escola**, n. 10, 43-49, Nov. 1999.

GUERREIRO, L. **Heavy metals in the body.** Available at: <http://alimentacaoviva.blogspot.com.br/2007/07/metais-pesados-no-organism> Accessed on March 05, 2013.

LEITE, R.F.; RODRIGUES, M. A. Environmental Education: Reflections on the practice of a group of chemistry teachers. **Ciência & Educaçao**, v. 17, n. 1, p. 145-161, 2011.

LEME, T. N. **The practical knowledge produced by teachers who do Environmental Education at school: traversing paths between theory and practice.** Dissertation (Master's in Education) - University of São Paulo, 2003.

LENARDÂO, E. J.; FREITAG, R. A.; DABDOUB, M. J.; BATISTA, A. C. F.; SILVEIRA, C. C. "Green Chemistry": the 12 principles of Green Chemistry and their inclusion in teaching and research activities. **Quimica Nova,** v. 26, n. 1, p. 123-129, 2003.

LIMA, J. D. de. **Sistemas Integrados** de **Destinaçao Final** de **Residuos**

Sólidos Urbanos. Paraiba: Gràfica JB, 2005.

LIMA, P. C.G.; LIMA, V. A.; KATAOKA, A. M. Overview Management Chemical Residues of Laboratories in Academic Institutions in Brazil. In: BRONIEWICZ, E. **Environmental Management In Practice.** Croatia: InTech, 2011. Pag. 351.

MACEDO, J. A. B. **Introduçao à Quimica Ambiental: Quimica & Meio Ambiente & Sociedade.** Minas Gerais: CRQ-MG, 2006.

MANO E. B.; PACHECO, É. B. A. V.; BONELLI. C. M. C. **Meio ambiente, poluiçao e reciclagem.** Sâo Paulo: Blucher, 2010.

MAROUN, C.A. **Manual de Gerenciamento de Residuos: Guia de procedimento passo a passo.** Rio de Janeiro: GMA, 2006.

MICARONI, R. C. C. M.; **Waste Management in Laboratories of the Institute of Chemistry of UNICAMP.** State University of Campinas: Doctoral Thesis - Institute of Chemistry, 2002.

MOZETO, A. A.; JARDIM, W. F. Environmental Chemistry in Brazil. **Quimica Nova**, v. 25, Supl. 1, p. 7-11, 2002.

PRADO, A. G. S. Quimica Verde, os Desafios Da Quimica Do Novo Milênio. **Quimica Nova**, v. 26, n. 5, p. 738-744, 2003.

SARI, R. SOTILES, A. R. LOSS, E. M. S.; VARELA, E. R. & ROCHA, R. D. C. Treatment of chromium (VI) waste generated in the Chemistry Laboratory. **Synergismus Scyentifica UTFPR**, v.10, n. 1, p.152-158, 2015.

SATO, M.; CARVALHO, I. C. M. **Educaçao Ambiental: pesquisa e desafios.** Porto Alegre: Artmed, 2005.

SILVA, F. M.; LACERDA, P. S B.; JUNIOR, J. J. **Sustainable Development and Green Chemistry.** Quimica Nova, Vol. 28, No. 1, 103-110, 2005.

SILVA, R. R.; MACHADO, P. F. L. Experimentation in high school chemistry: the necessary search for ethical-environmental awareness in the use and disposal of chemical products - a case study. **Ciência & Educaçao**, v. 14, n.

2, p. 233-249, 2008.

SISSINO, C. L. S.; OLIVEIRA, R. M. **Residuos sólidos, Ambiente e Saùde: uma viso multidisciplinar.** Rio de Janeiro: Editora Fiocruz, 2000.

SOTILES, A. R.; SILVA, L. D.; LOSS, E. M. S.; FERRI, E. A. V. & SARI, R. Determination of the purity of copper sulfate recovered from practical chemistry classes. **Synergismus Scyentifica UTFPR**, v. 10, n.1, p. 37-43, 2015.

TEIXEIRA, R. A. & RIBEIRO, A. P. **Actions to minimize waste production in practical chemistry classes in laboratories of a higher education institution.** IV International Symposium on Project Management, Innovation and Sustainability, Sâo Paulo, Brazil, 2015.

UNICENTRO. **General and Experimental Chemistry II course handout.** Guarapuava, 2012.

UNIFAL - Federal University of Alfenas. **Waste disposal**. Chemical Risks Committee. Available at: <http://www.unifal-mg.edu.br/riscosquimicos/descarte>. Accessed on: July 18, 2014.

UTFPR. **High School General Chemistry Laboratory Practices Workbook,** Pato Branco, 2013.

MIX
Papier aus verantwortungsvollen Quellen
Paper from responsible sources
FSC® C105338

Printed by Books on Demand GmbH, Norderstedt / Germany